Corporeality of Antigravity
Volume One

Corporeality of Antigravity
Volume One

An Antigravity Force, That Might Suddenly Become Incadescent In The Mind, Radiating Outward With Such Apocalyptic Power That Everything Would Change

Steven L. Basic
Senior Aerospace Stress Engineer (Ret.)
Boeing Corporation

Copyright © 2018 by Steven L. Basic.

ISBN:	Softcover	978-1-9845-6185-5
	eBook	978-1-9845-6184-8

All rights reserved. No part of this book may be reproduced or transmitted in any form or by any means, electronic or mechanical, including photocopying, recording, or by any information storage and retrieval system, without permission in writing from the copyright owner.

The views expressed in this work are solely those of the author and do not necessarily reflect the views of the publisher, and the publisher hereby disclaims any responsibility for them.

Any people depicted in stock imagery provided by Getty Images are models, and such images are being used for illustrative purposes only.
Certain stock imagery © Getty Images.

Print information available on the last page.

Rev. date: 10/27/2018

To order additional copies of this book, contact:
Xlibris
1-888-795-4274
www.Xlibris.com
Orders@Xlibris.com
787040

Contents

Foreword ix

1.) Abstract 1

2.) Antigravitational Repulsion Of The Sun 3

3.) Description Of Variables Within T1 Table Data 4

4.) An Elementary Calculation With The Planet Earth
    Data—Row, As In Tables T1 And T2 8

5.) Newton's Attraction Minus Negative Eternal Weight
    Equals To The Centrifugal Force 10

6.) The "Great Inequality Of Jupiter And Saturn" 12

7.) Ambient Power When A Reserch Laboratory Is Located
    On The Surface Of Planet Earth 15

8.) The Thesis Prominence 19

9.) A Definition Of The Antigravity Energy 21

10.) Conclusions 22

11.) Solid Core Sizes Of Outer Planets 24

12.) An Elementarty Theory Of A Previously Unknown
    Energy Of Nature 29

13.) Accelerated Rotation 32

14.) Total Frequency..34

15.) Stereo Radian Example...36

16.) Scalar Sum Of The Antigravity, The Classical, And The
High Energy Rotations..38

17.) The Third Frequency Of Galileo Galilei.................................40

18.) Figure 3..42

19.) Antigravity Energy Variance For All The Solar Planets,
Exhibiting Three Maximums, In Light Of The Observed
Activity At The Sun 43

20.) Tesla Ambient Energy 45

21) Addendum 49

22.) References..56

To my children:

Daughter, Adrienna Faith

and

Son, Christopher Michael

Corporeality of Antigravity

Foreword

The Universe is restless and in turbulent motion forcing its way towards balance and symmetry. Understanding the forces beyond the vast distances takes enormous amount of intellectual effort. For centuries, along with the knowledge accumulating on gravity, the existence of Antigravity occupied varying but cautious levels of attention over time.

Numerous respectable sources keep open their interest on what is Antigravity and where it came from. An anecdotal but true story of Newton's encounter with his contemporary the Reverend-Philosopher Robert Bentley, who forwarded to Isaac Newton a question in essence that due to gravity shouldn't the universe collapse? Being a human, confronted with at that time unanswerable question, in response, Newton calls on God. More specifically, the records tell us that Bentley wrote to Newton, "if all the stars and particles attract one another why the universe did not collapse?" The event remains dubbed the Bentley Paradox.

In the book *Corporeality of Antigravity*, the principal author Steven L. Basic uses mathematical analysis to cement the theories essential to understanding our Universe. Basic's work show that antigravity is no longer an idea but a defined component of the Universe. High-level scientists having studied the knowledge mounting during remarkable historic times, approve the importance of advanced mathematics accompanying the evolving scientific thought. With his work ***Corporeality of Antigravity***, the author S.L. Basic strongly confirms such approach. This book contains evidence and confirms that one thousand to one million times less intense than Newtonian interaction, does exist from which the power of the gravitational field can be extracted.

Motto: "If at first an idea does not sound absurd, then there is no hope for it." Albert Einstein.

Many noted scientists have emerged in human history. Newton (Sir Isaac, 1642-1727), served the foundations on classical mechanics and their applications to the motions of bodies in space. Famously, Isaac Newton defined the three Laws of Motion, being: First law on the principle of Inertia; the Second quantifies the concept of Force; and the Third law pointing to that to every action there is an equal and opposite reaction. We may assume that if there is a Gravity, and there is a gravity (the Action), there must be an Antigravity (the Reaction) preventing collapse of the universe. The astronomers insisted in their times, that the universe was static and unchanging. According to modern authorities (Ref. Kaku, page 293), Einstein approved the astronomical observations disclosing the cosmological constant, an antigravity force that pushed the stars apart to balance the gravitational pull; that this antigravitational force corresponds to the energy contained in vacuum. What is driving the current acceleration of the universe is the dark energy (Ref. Kaku page 291), which in turn is probably caused by the "cosmological constant."

The key is to understand the mysterious cosmological constant, or the energy of the vacuum. The WMAP satellite orbiting the Earth shows that this cosmological constant seems to be driving the current acceleration of the universe, but we do not know if it is permanent or not.

Recently, observations of supernovae have led most theorists to re-introduce λ or similar terms, referred to as "dark energy" (Ref. Penrose, page 463).

The name Tesla evokes electricity, and very strongly a postulate on the energy of vacuum. Tesla, Nikola (1856-1943) born Serbian national on Austro-Hungarian territory; studied in Budapest, and Vienna, and became highly recognized U.S. scientist. Among many, an important concept of the Energy of Vacuum is attributed to Tesla. Tesla's interests; the disclosures were exceeded by his own intellectual abilities.

For a sophisticated scientist it may sound simplistic to quote Le Chatelier. However, the science teaches us of a principle (Le Chatelier) that,

If a system in a balanced or equilibrium state is disturbed, the system will readjust in such a way as to neutralize the disturbance and restore equilibrium.

The Human thinking capacity applies all available knowledge; and this principle may well help explain the works of forces reaching beyond matter and into the antimatter.

Einstein, (Albert, 1879-1955) physicist, was born in Germany, and famously became adopted U.S. scientist. In 1916 Einstein stumbled upon Robert Bentley Paradox. (Ref. Kaku., page 292). Consequently, Einstein introduced the term cosmological constant. This is an exceedingly tiny constant quantity λ, whose actual presence is strongly suggested by modern cosmological observation. Very recently, observations of supernovae have led most theories to recall the importance of λ, therefore to reconfirm the dark energy. Einstein's *Lambda (λ), initially referred to as the "cosmological constant", implied to the existence of a repulsive form of gravity, the driving force behind cosmic acceleration. Today, the science defines this force as the **Dark Energy**. It has been said, "if dark matter gives rise to the gravity that holds the universe together, then dark energy is the counter-force pushing the universe apart." An alternative theory proposes that the universe may be filled with an exotic "quintessence" substance evolving from the dark energy. [Ref. Penrose page 777]. In continuation, a theory proposed that the dark energy could be due to neutrinos interacting with hypothetical particles called "accelerons." The faith of the universe rests on whether the dark energy is constant or changing. All the observations to-date indicate that the dark energy is constant. To summarize, Einstein's term cosmological constant: is equated with dark energy; is equated with the vacuum energy, or sometimes the quintessence. Today's state of knowledge holds that Lambda (λ) cannot change with time; and the constancy of λ is a direct consequence of the energy conservation equation.

Dirac, (Paul. 1902 – 1984). In 1933 Dirac won the Nobel Prize for his postulate predicting the antielectron. (Ref. Kaku page 188). Dirac's mathematical equation has two types of solutions: one for matter, and one on antimatter. Stephen Howking offered regret about Dirac not patenting his equation. Dirac's famous equation for the electron upgraded Einstein's equation to $E = +/-mc^2$. Dirac based his equation for the electron on his ability to use mathematical "spinors". Following such substantial thinking, noted authority on physics (Ref. Kaku page 190) had stated that "gravitation is its own antimatter".

Feynman, Richard, had revealed the true secret of antimatter: it is an ordinary matter going backward in time (Ref. Kaku, page 278). The puzzle is presented as the particles travel backward in time presenting them-selves as the antimatter.

Antigravity is described as an idea of creating a place or object that is free from the force of gravity. Newton's law of Universal gravitation states, "gravity was an external force transmitted by unknown means". In the 20th century, Newton's model was replaced by general relativity. Under general relativity, anti-gravity is impossible except under "contrived circumstances". Quantum physicists have postulated the existence of gravitons, mass-less elementary particles that transmit gravitational force, but possibility of creating or destroying these is unclear. "Anti-gravity" is often used to refer to devices that look as if they reverse gravity even though they operate through other means, such as lifters, which fly in the air by moving air with electromagnetic fields.

The nature of the dark matter is still not agreed upon by astronomers. (Re. Penrose, page 773). According to recognized authorities (Ref. Kaku, page 205), the Dark Matter is a mysterious substance that is invisible but has a weight; it is "perhaps ten times as plentiful as ordinary visible matter in the universe."

Negative energy is different from negative matter in that it actually exists in small quantities. (Ref. Kaku, page 206)

The scientists dedicated to physics point out to Hendrik Casimir, who in 1933 claimed that the vacuum between two parallel plates is not empty but contain virtual particles--which dart in and out of existence. Furthermore, because of uncertainty principle, that the tiny particles may be short lived, and the energy still obey the law of conservation of energy. The Casimir energy is said to be proportional to the inverse fourth power of the distance of separation between the plates.

In modern times a vast array of particles was scientifically evaluated and new particles discovered and disclosed in the public domain: the photon; electron; proton; positron; neutron; neutrino; muon; pion; kaon; lambda and sigma particles, and the famously predicted omega-particle.

The antiproton was directly observed in 1955, and the antineutron in was observed in 1956. There are new kinds of entity known as quarks, gluons, and W and Z bosons. There are vast groups of particles referred to as "Resonances". The modern theory also demands transient entities called "virtual" particles; and quantities known as "ghosts", far from observability.

For many years during his residences in Cambridge, United Kingdom, as well generously adopted U.K. and U.S. scientist and engineer, Steven L. Basic have studied tirelessly many above mentioned historical personalities, to the point that he understood their place in history.

The author Steven L. Basic studied the works of Galileo Galilei, Kepler and Copernicus.

The history discloses an ability of the scientists to observe, analyze by thinking, and apply the mathematics skills. To prove a theory the mathematics are applied. For example, Kepler's Laws are derivable from Newton's Laws of Motion; and Newton's Laws of gravity are derivable from Kepler's Laws.

As well Basic studied the theories of Heisenberg and Schrodinger. As well, the author Basic has explored and used astronomical data firmly established on the Solar System leading planets. Beyond expectations, in author Basic's unique view the Zero Point Energy has re-emerged from an unexpected direction, and randomly renamed to the Low Energy Nuclear Reactions.

Terminology

Antigravity [Phys.] Ref. McGraw Hill, page 78. The repulsion of one body by another by means of gravitational-type of force.

Gravity [Mech.] Ref. McGraw Hill, page 638. The gravitational attraction at the surface or other celestial body.

Graviton [Phys.] Ref. McGraw Hill, page 638. A theoretically deduced particle postulated as the quantum of the gravitational field, having rest mass and charge of zero, and a spin of two.

Cosmological constant [Relat.] Ref. McGraw Hill, page 336. The multiplicative constant for a term proportional to the metric in Einstein's equation relating the curvature of space to the energy-momentum tensor. Cosmological constant is the simplest possible form of dark energy since it is constant in both space and time.

Gravitational Systems of Units. [Mech.] Ref. McGraw Hill, page 638. System in which length, force, and time are regarded as fundamental, and the unit of force is the gravitational force on a standard body at a specified location on the earth's surface.

Dark Energy, is the energy of the vacuum. The Dark energy comprise73% of the universe. ***Dark Matter***, comprise about 27% of the universe.

Foreword by: Zora L. Toth, Ph.D. Research Chemist. Member Emeritus of the American Chemical Society.

References

McGraw Hill. Dictionary of Scientific and Technical Terms. @1974

Kaku, Michio. *Physics of the Impossible*. Doubleday. @ 2008

Title: *Einstein's Biggest Blunder*. APS News July 2005 (Volume 14, Number 7)

Penrose, Roger. *The Road to Reality*, Understanding of our Physical World. @2004

Ferreira, Pedro G. *The Perfect Theory*, A Century of Geniuses and Battle Over General Relativity. Houghton-Miffin-Harcourt. @ 2014.

1.) Abstract

The main objective of the book is to provide a Corporeal evidence that the "…RADIATING OUTWARD…" Force does already exist within the Solar system; its visible effect has been observed for many Centuries as the "GREAT INEQUALITY OF JUPITER AND SATURN". The celebrated Swiss Mathematician Leonhard Euler had submitted to the French Academy of Sciences a monograph concerning this Natural Phenomenon.

The Antigravity force in Nature, has a negative polarity, i.e., it is acting in a direction opposite to the Newton's Attraction, or, it is repulsive.

Newton's Attraction predominates in the case of all inner planets: Mercury, Venus, Earth, Mars and Jupiter. However, in the case of all outer planets: Saturn, Uranus, Neptune and Pluto, this "NEGATIVELY POLARIZED", repulsive, "ANTIGRAVITY", previously unknown force, somewhat reduces the Newton's Attraction and causes the outer planets to move towards the Outer Universe.

In essence, if and only if, the Hyperbolic Potential Energy of Sir Isaac Newton would exist alone, the forces within the Solar System would be simplified. The fact is that in real Nature the Great Inequality of Jupiter and Saturn does exist, and the existence of the ANTIGRAVITY causes this disturbance. According to Pythagoras: "Numbers Take Man by the

Hand and Conduct Him Unerringly Along the Path of Reason." The numbers presented in TABLES T1, and T2 prove that:

[1] Ambient energies are proportional to the intensities of the Planetary surface gravitational fields "g".

[2] The intensities of the Planetary surface gravitational fields "g" are proportional to the Zero Point Energies of Einstein—Stern.

[3] The intensities of the Planetary surface gravitational fields "g" are proportional to the stereo radians that each Planet occupies within the Euclidean Solar space.

[4] The intensities of the Planetary surface gravitational fields "g" are proportional to the velocity of light in Outer Space, squared.

A Dynamic--Static equilibrium of the following forces has been considered in this book:

[1] The centrifugal force, acting upon Planetary mass, in outward direction

[2] Newtons Attraction: attracting Planetary masses toward the center of the Sun.

[3] Antigravitational repulsion of the Sun, acting opposite to the gravitational attraction.

2.) Antigravitational Repulsion Of The Sun

Statement

The results, obtained by means of mathematical--analytical analysis has been shown in the attached Tables "T1" and "T2" and "T3":

From nearly total equality between the columns eight (8) and nine (9) in table T2, Page number 7, the following Theorem may be deduced:

The "Ambient Energies" of all Solar planets, that are in a state of relative rotation, are proportional to the planetary surface gravities "g"; also, to the ZERO POINT ENERGY of Einstein—Stern. The ambient Energy has a wave--form and emanates from the center of the Sun. Each Planet absorbs the gravity waves in proportion to its Planetary Stereo Radian.

3.) Description Of Variables Within T1 Table Data

Column [2]: are Total angular velocities of Planets as a simple Scalar sum between the angular, positive or negative, velocity in respect to the planetary polar axis, plus the angular velocity in respect to the center of the Sun.

This Scalar sum is a total angular velocity of a relative rotation. [radians per second].

Column [3]: are planetary frequencies, or, Column number 2 divided by (2*3.14159…) [Hertz]

Column [4]: are outer planetary radii [Meters]

Column [5]: are distances of the Planetary orbits from center of the Sun.

Column [6]: are the "Ambient Energies" = E A

$$E\,A = C\verb|^|2 * \sqrt{\left(f * \frac{h}{2}\right) * \left(\frac{r}{R}\right)^2}$$

Where: "C": is the velocity of light in outer space [Meters/sec]. "f": is the frequency [Hertz] "h": is the constant of Max Planck [In "MKS" system h=6.4942*10^-33 Kilogram*Meter*Second]

"$\left(\dfrac{r}{R}\right)^2$": are Stereo Radians in the Three-Dimensional Space of Euclid.

"$\left(f * \dfrac{h}{2}\right)$" : are the Zero Point Energies of Albert Einstein and Otto Stern [Kilogram*Meter].

TABLE T1

(1)	(2)	(3)	(4)	(5)	(6)
Planet	Total angular velocity [Rad/sec]	Total frequency [2]/ (2*3.14159) *10^--5 [1/sec]	[r] *10^6 [Met.]	[R] *10^11 [Met.]	$+/-\sqrt{\left(f*\frac{h}{2}\right)*\left(\frac{r}{R}\right)^2} = +/-\sqrt{\left([3]*\frac{h}{2}\right)*\left(\frac{[4]}{[5]}\right)^2}$
Mercury					
Venus					
Earth	7.312*10^--5	1.164*10^--5	6.37	1.49	8.309*10^--24
Mars	7.098*10^--5	1.130*10^--5	3.32	2.28	2.789*10^--24
Asteroid					
Jupiter	1.773*10^--4	2.823*10^--5	69.8	7.78	2.716*10^--23
Saturn	1.705*10^--4	2.714*10^--5	59.2	14.3	1.228*10^--23
Uranus					
Neptune					
Pluto					

TABLE T2

(1) Planet	(7) [6] *(C)^2	(8) [7] *12.228*10^6 CALCULATED SURFACE GRAVITATION [Meters/(sec)^2]	(9) "OBSERVED" NEWTONIAN PLANETARY SURFACE GRAVITATION [Meters/(sec)^2]
Mercury			
Venus			
Earth	7.476*10^--7	9.144	9.81
Mars	2.509*10^--7	3.068	3.727
Asteroid			
Jupiter	2.443*10^--6	29.873	25.928
Saturn	1.1055*10^--6	13.518	11.369
Uranus			
Neptune			
Pluto			

4.) An Elementary Calculation With The Planet Earth Data— Row, As In Tables T1 And T2

THE EXPRESSION TO BE CALCULATED:

$$-g = C^2 {}_{*}\left\{+/-\sqrt{\left(f*\frac{h}{2}\right)*\left(\frac{r}{R}\right)^2}\right\}12.227*10^6$$

[Meter/sec^2] ....2.)

Independent variables of this expression are (from the left-hand side):

"--g": is the intensity of the Earth's surface acceleration selecting the Negative square root value. [Meters/sec^2].

"(f*h/2)": is the "Zero Point Energy" of Einstein—Stern [Kilogram*meters].

"f": is the total frequency [Hertz], in the case of the Planet Earth f = 1.164*10^--5.

"h": is the constant of Max Planck [In "MKS" system h=6.49428*10^--33 Kilogram*meter*sec].

"r": is the outer radius of the Planet Earth [= 6.37*10^6 Meters].

"R": is the distance between the orbit of the Planet Earth and the center of the Sun [= 1.49*10^11 Meters].

"(r/R) ^2": is the "Stereo Radian" in Euclidean Space [Dimensionless].

"12.227*10^6": is the conversion coefficient to the "MKS" system.

"C^2" = 8.9979*10^16 [Meters^2/sec^2].

Substituting these values into the expression No. 2:

8.9979*10^16*{+/--Square root [(1.164*10^--5/2*

*6.4942*10^--33) *(6.37*10^6/1.49*10^11) ^2]} *12.227*10^6 =-- 9.144 [Meters/sec^2]

This value has been found within the Table T2, with positive square root solution.

5.) Newton's Attraction Minus Negative Eternal Weight Equals To The Centrifugal Force

When the negative square root solution surface gravitation is multiplied with the "mass—zero", it becomes an antimatter image of the Eternal Weight. It will be anticipated, initially, that the Newton's Attraction will predominate, it does, only with relatively not known, hypothetical, radii of Outer Planets Jupiter, Saturn, Neptune and Pluto.

Therefore, the Centrifugal Force acts Outward and the Newton's Attraction acts in opposite direction, towards center of the Sun, however, this initially assumed to be smaller than the Newton's attraction, antigravity Force acts in opposite direction—it has a repulsive action:

$$(m\ planet) * \frac{V^2}{R} =$$

$$\frac{(m\ planet) * (m\ sun) * (Gamma)}{R^2} -$$

$$-(m\ planet) * g \dotfill 3.)$$

Hence, cancelling (m planet) at both sides and transferring "R" to the right-hand side, expression 3 becomes, according to Joseph Lagrange, "Vis Viva":

$$V^2 = \frac{(m\ sun) * (Gamma)}{R} -$$

$-g * r$ (**Refer to the expression 14.**)......................4.)

The first term at the right-hand side is the Newton's attractive field potential. However, the second term is an antigravity energy of nineteenth Century assertion of Marie Curie; therefore, this term will be referred to as the antigravity energy of Marie Curie. Its single factor (r/R) (for Solar Planets) follows from The Thesis Prominence, below on Page 18.

6.) The "Great Inequality Of Jupiter And Saturn"

The square root of the expression number 4 will be calculated:

$$V = \sqrt{\frac{(m\ sun)*(Gamma)}{R} - g*r} \quad \text{............5.)}$$

This expression No.5 asserts, conservatively, that the energy of Madame Marie Curie = g*r is non—oscillating, non--radiating from the center of the Sun, nor absorbed by Planets per Stereo—Radian.

A more realistic form of the expression No. 5.), or, of the second term, per Stereo Radian, would read:

$$(g*r)*\left(\frac{r}{R}\right)^2 = \frac{g*(r^3)}{R}$$

From center of the Sun.
This more realistic expression will be considered in detail within the "Corporeality of Antigravity ", Volume 2.

for all the Solar Planets and shown within the last column No. 6, of the Table T3, page number 13. Surprisingly all the calculated orbital tangential velocities in respect to the center of the Sun are in good agreement with the observations for all the Planets: Mercury. Venus, Earth and Mars.

In the case of outer Solar Planets, Jupiter, Saturn, Uranus, Neptune and Pluto, however, the tangential velocities are a Complex number, since the inner solid cores sizes of outer Planets are NOT very well known and the second negative antigravity term predominates; and this would confirm the observed fact that the outer Planets velocities, may be Complex, causing the Saturn, Uranus Neptune and Pluto, to move away and once in the distant future may altogether be lost from the Solar System. This possibility is shown on the Page 17, Fig. 1. It may be of some interest to note that the calculated tangential velocities of inner Planets are somewhat (infinitesimally) smaller than observed, because of the existence of the antigravity (refer to TABLE 3)

TABLE T3

(1) Planet	(2) ORBITAL RADIUS FROM CENTER OF THE SUN "R" [Meters] *10^11	(3) OBSERVED SUFACE GRAVITY "g" [Met./Sec^2]	(4) PREDOMINATING NEWTON'S ATTRACTION POTENTIAL 1.322*10^20/R *[10^6]	(5) ANTIGRAVITY FROM THE SUN IN LIGHT OF MADAME MARIE CURIE ASSERTION {[3] *[4, Table 1] IN [MKS] SYSTEM *10^6	(6) SQUARE ROOT FROM [(4)— (5)] [Met./sec]
Mercury	0.579	3.728	2283.242	--8.723	47692 Below observed
Venus	1.08	8.86	1240.074	--55.237	34421 Below
Earth	1.49	9.81	887.248	--62.489	28719 Below
Mars	2.28	3.727	579.824	--12.373	23821 Below
Asteroid					
Jupiter	14.3	25.928	169.922	--413.774 Complex	Solid core??
Saturn	14.3	11.369	92.447	--673.044 Complex	Solid core??
Uranus	28.7	10.89	46.062	--259.619 Complex	Solid core??
Neptune	45	11.87	29.377	--265.888 Complex	Solid core??
Pluto	59.1	4.218	22.367	--12.654 Complex	Solid core??

7.) Ambient Power When A Reserch Laboratory Is Located On The Surface Of Planet Earth

When a Research Laboratory is located on the surface of planet Earth, the Stereo Radian in respect to the Sun, remains the same and, according to the Theory of Relativity, the following expression may apply:

> The square of the ambient energy REST MASS is simply proportional to the Zero Point Energy of Einstein – Stern.

Referring to the page 8 in this book, the Expression 2.),

$$--g = C^2 * \{+/- \sqrt{\left(f * \frac{h}{2}\right) * \left(\frac{r}{R}\right)^2}\} * 12.227 * 10^6$$

[Meter/sec^2] .... 6.)

Where:

"r": is the outer radius of the Planet Earth [= 6.37*10^6 Meters].

"R": is the distance between the orbit of the Planet Earth and the center of the Sun [= 1.49*10^11 Meters].

"(r/R)^2": is the "Stereo Radian" in Euclidean Space [dimensionless].

"[12.227*10^6": is the conversion coefficient to the "MKS" system.

Substituting these terms into the Expression number 1, the negative gravity will become

$$--g = C^2 * \{+/- \sqrt{\left(f * \frac{h}{2}\right) * \left(\frac{6.37*10^6}{1.49*10^{11}}\right)^2}\} * 12.227*10^6$$

....2.)

A noted contemporary American authority on physics (Ref. Kaku, page 190) had stated to the effect that "Particles that have no charge ... can be their own antiparticle". Accordingly, we may state that "GRAVITATION IS ITS OWN ANTIMATTER".

$$--g = C^2 * \{+/- \sqrt{\left(f * \frac{h}{2}\right)} *\} * 522.704$$

....3.)

This, Expression 3.) is the "First Postulate" by Steven L. Basic in his previously published book "UNIFIED THEORY" (2009).

The term under the square root (f*h/2) is the Zero Point Energy of Einstein—Stern.

From the expression 3.), the following Theorem may be obtained:

The square of the Ambient Energy ("—g/C^2") ^2 is SIMPLY PRPORTIONAL TO THE ZERO POINT ENERGY of Einstein—Stern.

With this Theorem the Extended Theory of the "LOW ENERGY NUCLEAR REACTIONS" begins.

$[(-g)/(C^2)]^2 = h*f*136642.56$ (From Equation No. 2) ....7.)

Hence: $(-g)^2 = C^4*h*f*136642.56$ ....8.)

The values of "C" and "h" in metric "MKS" system, are:

"C" = $2.9979*10^8$ [Meters per second]

"C^4" = $8.0773*10^{33}$ [Meters per second] ^{4}

Max Plank constant "h" = $6.4942*10^{-33}$ [Kg force*sec]

Substituting this data and calculating the Expression No. 8.), the negative gravity will be: $(-g) = 2677*\sqrt{f}$ .... 9.)

By multiplying both sides of the equation No. 9.)
with the mass of the oscillating matter and "r"

E ANTIGRAVITY = $(--m*g*r)$ (Madame Marie Curie) ....9.1.)

From $--g = -*2677*\sqrt{f}$ Kg*sec^2/m, in 9.1), f = $(9.81/2677)^2$ = 1.932&10*--5 (Observed Total frequency of the Earth......10.)

Conditionally antigravitation work done due to the Eternal weight: $(--g*m*r)$, acting upon a finite displacement 'r', may be understood as the only Ambient Energy that is available, than this Energy is proportional to the Power of the System (Internet, Max Planck) and the Ambient Power will read: P = $(m*g*r)/75/1.37$ [Kilowatts] .... 11.)

(where "r" is the distance between the centroid of the
oscillating weight and the axis of the oscillations)

The expression No. 11.), applied to the Existing--Tested, Machines for extraction of Power from the gravitational Field [References No. 1.) and No. 2.)] leads toward a reasonable Energy = Power Levels.

Figure 1

Newton's Attraction and Antigravity of Madame Marie Curie

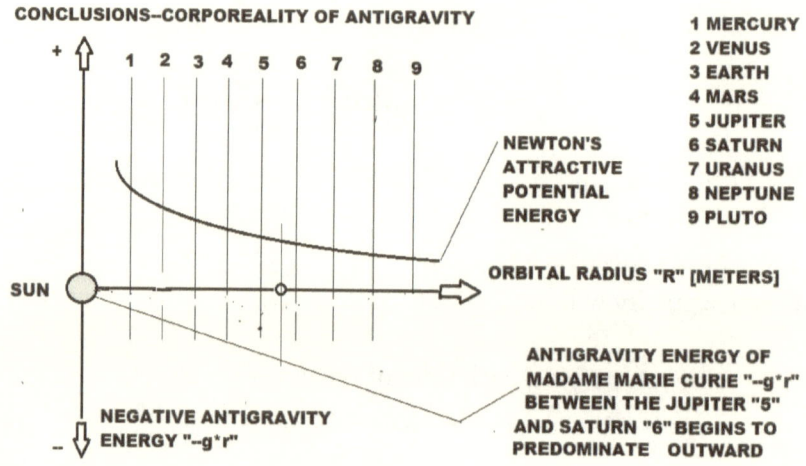

8.) The Thesis Prominence

"Even when the Nature of the space becomes better known, the cause of the reality will remain a mystery and the phenomena dependent upon the extent will most likely always present itself to us as a profound and wonderful enigma".

With reference to this book, expression no. 2 on page 8, the negative "g":

$$-g = (C^2)\left\{\left(\genfrac{}{}{0pt}{}{+}{-}\right)\sqrt{\left[\frac{(h)*(f)}{2}\right]*\left(\frac{r}{R}\right)^{\wedge}2}\right\}*12.227*10^{\wedge}6 \quad ......2$$

The stereo radian: "(r/R) ^2" is equal to "1",

because "r" is equal to "R" at the surface of the Sun, and the initial boundary condition will be:

$$-g = (C^2)*\left[\left(\genfrac{}{}{0pt}{}{+}{-}\right)*\sqrt{\frac{(f*h)}{2}*(1)}\right]*(12.227*10^{\wedge}6)....12.)$$

To upgrade this initial boundary condition 12.) to be valid for all the Solar Planets, both sides of the expression No. 12) will be multiplied with the square root of a Stereo Radian:

$$\sqrt{Stereo\ Radian} = \sqrt{\left(\frac{r}{R}\right)^2} \quad ....13.)$$

$$-g * \left(\frac{r}{R}\right) = (C^2) * \left[\left(\frac{+}{-}\right) * \sqrt{\frac{(f*h)}{2} * \left(\frac{r}{R}\right)^2}\right] * 12.227 * 10^6 \dots 14$$

By multiplying the left-hand side of the expression no.14 with the Planetary Orbital distance from the center of the Sun "R" (refer to the expression no. 4,

(--g*r/R) *R= (--g observed) r.......................................15.)

Where "r" is the extent, or the radius of the rotating, or, oscillating mass-like event, and "g observed is the observed surface gravitation.

The expression no. 15. will be denoted "The Antigravity Energy of Madame Marie Curie", asserted by her during defense of the post--graduate work, at the end of 19th.

9.) A Definition Of The Antigravity Energy

The Antigravity Energy has a negative polarization relative to the positive Newton's attraction potential. The energy has been famously asserted by Marie Curie, therefore will be denoted as E_{MC}:

$$E_{MC} = (--g^* m_0)^* r \ [(kg\ Force) * meter] \ \ldots\ldots 16.)$$

Where: "g" is the surface gravitation of the orbiting – satellite rest-mass m_0; $(--g^* m_0)$ is the negatively polarized Eternal weight and "r" is the extent, or the radius of the orbiting satellite rest mass m_0.

10.) Conclusions

Uniform Radiation from the Sun

From the point located between Planets Jupiter, in Dia. No. 1, denoted with the number "5", and the Planet Saturn "6", toward Outer Universe, the Uniform Antigravity potential would predominate, therefore, as observed, outer planets Saturn, Uranus, Neptune and Pluto, would be increasing their orbital distances from the center of the Sun. This Natural phenomenon is known for many centuries as "The Great Inequality of Jupiter and Saturn".

This Unique High Energy Mechanical Algorithm follows from the solutions of the differential equations with variable coefficients of celebrated Swiss mathematician, Professor Leonhard Euler and Professor Joseph Lagrange, and holds true in the case of Atomic Structures, and defines a previously not known Energy in Nature.

When a comparison between Observed Tangential velocities of Planets and the calculated Tangential velocities, is made, (column number 6, Table 3, of this book that takes in the account the action of antigravity), then it can be concluded that the calculated Tangential velocities of the Planets: Saturn, Uranus, and Neptune from the Table T3 are, even with reduced solid core radii, somewhat smaller, indicating that the antigravity, with its negative sign of the gravitational square root, is a Corporeal solution of the "Reverend – Philosopher Robert Bentley

PARADOX" for the Solar System, and the Outer Space. With a high degree of certainty, we may conclude that, THE GREATER THE GRAVITY AND THE CORE SIZE EXTENT (SIZE) OF A MASS— LIKE EVENT, THE GREATER WILL BE THE REPULSION, OR, THE ANTIGRAVITY.

Dia. 1

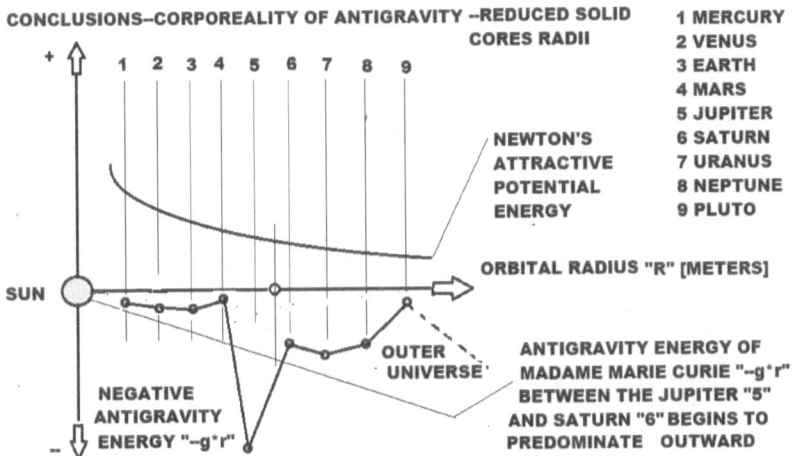

11.) Solid Core Sizes Of Outer Planets

In light of world-wide Encyclopedia sources, the exact solid core sizes of the outer planets (Jupiter, Saturn, Uranus, Neptune and Pluto), are NOT known yet at the beginning of the 20th century). As the interest on the subject persists, very many Space Exploration Establishments are attempting to measure the core sizes using a close fly-by radar equipment. Meanwhile, there are many assumptions in the contemporary public domain. The following assumptions have been applied in this study:

1.) The solid core size of the planet Jupiter = 1.5 times radius of the planet Earth = $1.5 * 6.37 * 10^6$ = 9,555,000 [meters]

2.) The solid core of the Planet Saturn = 1 times radius of the planet Earth = $6.37 * 10^6$ [meters]

3.) The solid core of the Planet Uranus = 1 times radius of the planet Earth = $6.37 * 10^6$ [meters]

4.) The solid core of the Planet Neptune = 1 times radius of the planet Earth = $6.37 * 10^6$ [meters]

5.) The solid core size of the planet Pluto remains known.

This set of assumption is shown as in column no "4" of the table T4, the page No. 25. The antigravities are calculated with this set off the reduced solid core sizes for Jupiter, Saturn, Uranus and Neptune and shown in column No. 5 of the table T5, page No. 26, and the diagram D 1, page No. 30.

TABLE T4

(1)	(4)
Planet	[r] [Met.]
Mercury	2,340,000
Venus	6,260,000
Earth	6,370,000
Mars	3,320,000
Asteroid	
Jupiter	9,550,000
Saturn	6,370,000
Uranus	6,370,000
Neptune	6,370,000
Pluto	3,000,000

TABLE T5

(1) Planet	(2) ORBITAL RADIUS FROM CENTER OF THE SUN "R" [Meters] *10^11	(3) OBSE--RVED SUFACE GRAVI-TY "g" [Met. / Sec^2]	(4) PREDOMI-NATING NEWTON'S ATTRACTION POTENTIAL 1.322*10^20/R *[10^6]	(5) ANTIGRAVITY FROM THE SUN IN LIGHT OF MADAME MARIE CURIE ASSERTION {[3] *[4, Table T4] IN [MKS] SYSTEM *10^6	(6) SQUARE ROOT FROM [(4)— (5)] [Met. /sec]
Mercury	0.579	3.728	2283.242	--8.723	47692 Below observed
Venus	1.08	8.86	1240.074	--55.237	34421 Below
Earth	1.49	9.81	887.248	--62.489	28719 Below
Mars	2.28	3.727	579.824	--12.373	23821 Below
Asteroid					
Jupiter	14.3	25.928	169.922 Higher	--247.612 Complex	Solid core??
Saturn	14.3	11.369	92.447 Higher	--72.427	4474 Below
Uranus	28.7	10.89	46.062 Higher	--69.369 Complex	Solid core??
Neptune	45	11.87	29.377 Higher	--75.611 Complex	Solid core??
Pluto	59.1	4.218	22.367 Higher	--12.654	3,116 Below

From the table T5, page No. 22, columns No. 5 and No. 6, below the Jupiter, towards the Outer Universe, the reduction toward solid core size, in the case of planet Saturn, has indicated a slightly favorable orbital tangential planetary velocity; however, orbital tangential planetary velocities of Jupiter, Uranus and Neptune, have remained somewhat complex numbers. Therefore, it would be reasonable to expect the exact solid core sizes, once they are to be determined, to be even smaller than those assumed by contemporary astronomers and applied (Refer to page No. 20) of this study.

12.) An Elementarty Theory Of A Previously Unknown Energy Of Nature

Antigravity "ENERGY", that had been asserted by Madame Marie Curie, for a special region of the moderate energies, is given by:

$$E_{MC} = --(mass)*(planetary\ radius)*(surface\ gravitation) = E_{MC} = --m_0*r*g$$

[(kg of force) *meters] ………………………………….. [17]

DEFINITION OF THE "ANTIGRAVITY" ENERGY

The Antigravity Energy has a negative polarization— relative to the positive Newton's attraction potential. This energy has been famously asserted by Madame Marie Curie, therefore will be denoted with E_{MC}:

$$E_{MC} = (--g*m_0)*r\ [(Kg\ Force)*meter] \ldots\ldots\ [18]$$

Where: "g" is the surface gravitation of the orbiting— satellite rest mass "m_0"; $(--g*m_0)$ is the negatively polarized Eternal weight and "r" is the solid core extent, or, the radius of the orbiting—satellite rest mass "m_0".

However, the Antigravity Energy incorporates the Surface gravitation "g" Wave and therefore is a Wave, according to the Internet Encyclopedias (and Max Planck); it is equal to the Wave Power:

$$(E_{MC}) = -(m*r*g) = \text{Power } [(Kg \text{ of force}) *(meters/second)] \quad \text{......... [19]}$$

For maximum power, an optimum Galileo Galilei's frequency has been found to be: (NON—QUANTUM)

$$f_{opt} = f_{GG3} = 0.407/\sqrt{R}[\text{Hertz}] \text{ is not very far}$$
from the frequency of Galileo Galilei:

$$f_{GALILEO\_GALILEI} = 1/6.28318*\sqrt{(g/R)} = .498/\sqrt{R} \text{ [Hertz]}.$$

.... [20]

And, since the Two Stage oscillator designed by Reference No. 1 (and similar gravity assisted machines) are operating NOT very far from both frequencies, the power for optimum frequency will be applied:

$$P_{OPT} = f_{GG3}*m*g*R = 0.407/\sqrt{R}*m*g*R \quad \text{......... [21]}$$

By multiplying and, simultaneously, dividing this expression by the square root of the extent, or, length of the pendulum, the maximum Power becomes:

P_{OPT} = (The Conversion Coefficient between Kg*m/second to Kilowatts) *(Oscillating Weight in kg of force) *(square root of the "extent", or, the second stage driven pendulum length)

The Conversion Coefficient between Kg*m/second and Kilowatts in this case = 0. 407/75*1.37 = 0.003961.. [22]

With this coefficient, the maximum Power becomes:

Popt = (0.003961) * (Oscillating Weight in kg of Force) *(square root of the "extent", or, the second stage, driven oscillator length) (this is the method for the ambient Power calculation) .... [23]

(NON—QUANTUM) Examples: Optimum Antigravity Powers:

Reference 1: A water pump both stages Powers added, P_{opt} = 0.14 Kilowatts,

Reference 2: P_{opt} = 75 Kilowatts.

Surprisingly, when the polarization sign is reversed from minus to plus the Antigravity potential, NON—QUANTUM upper limit of the Ambient Power Output, for oscillations of the two, or, a single stage oscillator of references No. 1 and No. 2, whose masses oscillate from "R" to "—R", will be obtained:

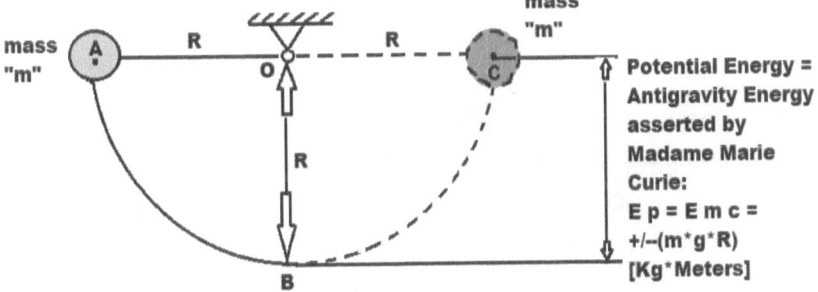

AMBIENT ENERGY OF THE GRAVITATIONAL STRESS FIELD

Potential Energy = Antigravity Energy asserted by Madame Marie Curie:
E p = E m c = +/–(m*g*R)
[Kg*Meters]

13.) Accelerated Rotation

Since the definition of the "antigravity" energy incorporates the magnitude of the Planetary surface gravitational acceleration "g":

the Antigravity Energy has a negative polarization relative to the positive Newton's attraction potential.

This energy has been famously asserted by Madame Marie Curie, therefore will be denoted with E_{MC}:

$$E_{MC} = (-\!-g*m_0)*r \ [(Kg \ Force)*meter] \ \ldots\ldots (24)$$

Where: "g" is the surface gravitation of the orbiting—satellite rest mass "m"; $(-\!-g*m_0)$ is the negatively polarized Eternal weight, and "r" is the solid core extent, or, the radius of the orbiting—satellite rest mass "m_0".

In this section, all the components of the magnitude of the Planetary surface acceleration "g" will be graph—analytically shown, to obtain essential conclusions about the nature of the Antigravity Energy.

With reference to the Page No. 8 of this book, the negative gravitation is given by:

$$-g = C^2 \cdot \{+/- \sqrt{((f \cdot h/2) \cdot (r/R)^2)}\} 12.227 \cdot 10^6 \; [\text{Meter/sec}^2] \quad \ldots\ldots\ldots\ldots (25)$$

It indicates the "Total Frequency" "f" as the first and the main independent variable.

14.) Total Frequency

The first angular velocity is in respect to the center of the Sun:

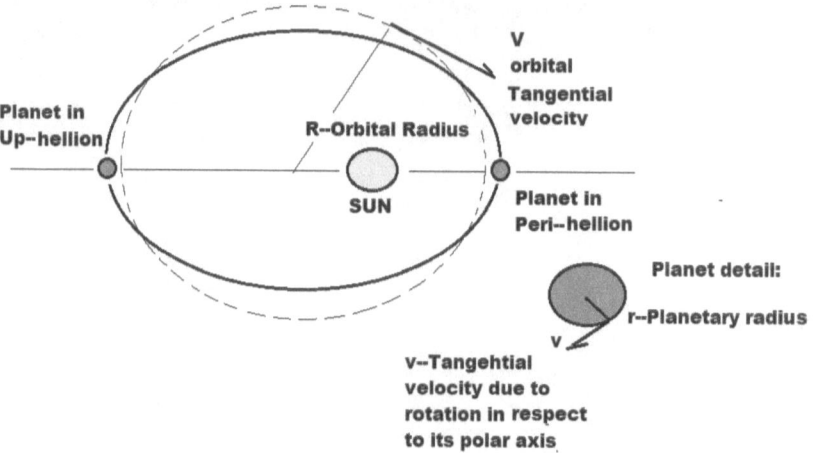

Fig. 2 Relative rotations of the Solar Planets

(Omega 1) = V/R [Rad/sec] [Eccentricity of the Sun is shown larger then reasonable] .... [25]

The second angular velocity is in respect to its polar axis:

(Omega 2) = v/r [Rad/sec] .... [26]

According to Classical and High Energy theory, the total angular velocity is equal to the Scalar sum:

(Omega total) = [(Omega 1) + (Omega 2)]

Hence, the (Total Frequency):

(Total Frequency) = [(Omega 1) + (Omega 2)]/ (2*3.14159) [Hertz] .... [27]

Consequently, the Total Frequency indicates Planetary passage from "Up--hellion" to "Peri—hellion", and again to "Up—hellion" = One Cycle of Rotation = Single Rotation.

Hence, the total frequency indicates that the Antigravity has the rotation feature.

15.) Stereo Radian Example

Example 1: you measure the light coming from a powerful globe. Your sensor is 50mm × 50mm in size, and if you hold it 2m away it measures 0.1 watts.

What is the radiant intensity in w/sr (watts per stereo radian)? answer: at 2m, one stereo radian cuts through 2×2 = 4 m² of the sphere, and because the sensor is relatively small; its flat surface area is approximately the area of sphere that it occupies. So, 0.05 × 0.05 = 0.0025m².

So, one stereo radian receives about 0.1 w × 4m²/0.0025m²) = 160 w/sr.

The radiant intensity of the light coming from the powerful globe, that is absorbed by the measuring sensor will be 0.1 watts:

160 watts/ (stereo radian) *((0.0025 m[^2])/ (4 m^2]))

(will be absorbed by the sensor) = 0.1 watts (at the sensor)

Just replace "measuring sensor" by "Planetary absorbing area" = (r^2) *3.14159.

If "R" is the distance of the orbit from the center of the sun, the absorbed power from the Sun will be proportional to (r^2)/(R^2) Stereo Radians

Conclusion:

Both components of the surface gravitational intensities, the Frequency "f" and the Stereo Radian "(r^2/R^2)" are indicating that the Nature of the surface gravitational Energy is Rotational; therefore, Antigravitational and NON--QUNTUM Energies, may come from a Scalar Sum of the Total Energy.

16.) Scalar Sum Of The Antigravity, The Classical, And The High Energy Rotations

E total = [Classical Rotational Energy – Rotational Antigravity Energy]

E total = [1/2*(m*R^2) *(Omega)^2--m*g*R] ... [28]

By taking the parametric variables, the mass "m" and the distance of the oscillating mass centroid from the axis of rotation "R" in front of the parentheses:

E total = m*R*[1/2*(R^2) *(Omega)^2--g] .... [29]

However, according to Classical and High Energy Theory: (Omega)^2 = (2*3.14159) ^2*f^2 .... [30]

Substituting this into the parentheses of E total, and dividing by "2":

E total = m*R*[19.739175*(R^) *f^2--g] .... [31]

Where "f" is the frequency [Hertz]

It is possible to conclude that the expression within parentheses will become zero, when:

$$f_{GG2} = (0.705)/\sqrt{R} \text{ [Hertz]}$$

This will be defined as the second frequency of GALILEO GALILEI "Galileo Galilei 2".

17.) The Third Frequency Of Galileo Galilei

The third frequency of GALILEO GALILEI "Galileo Galilei 3" is the frequency of the MAXIMUM AMBIENT Power abscissa,

that has been observed, at every initial stage
of an accelerated Rotation stage.

$$E \text{ total} = m*R*[1/2*(R^\wedge) *(Omega)^2 -- g] \dots [32]$$

When both sides of this expression for Energy, in light of Classical and High Energy Theory, are multiplied with the frequency "f", the Power of the System will be obtained:

$$P \text{ Ambient Power} = m*R*[1/2*R*(2*3.14159)^{\wedge}2*f^{\wedge}2*f - f*g] \dots [33]$$

Or:

$$P \text{ Ambient power} = m*R*[19.7391*R*f^{\wedge}3 -- f*g]$$

Equating the partial derivative of the "P Ambient Power" in respect to the frequency "f" and equating this derivative to zero:

$$[3*19.7391*R*f^{\wedge}2 -- g] = 0$$

The third frequency of GALILEO GALILEI "Galileo Galilei 3", will be obtained, corresponding to the Maximum Ambient Power:

$$f_{GG3} = (0.407)/\sqrt{R}$$

It may be of some interest to note that the Author, to prove the generality of this very unusual conclusion, has analytically determined the third frequency of Galileo Galilei for Planet Earth, whose radius is equal to $6.37*10^6$ Meters:

$$f_{GG3} = (0.407)/\sqrt{(6.37*10^6)} = 1.6125*10^{(-4)} \text{ [Hertz]}$$

This is, surprisingly, almost the same as the observed total frequency = $1.164*10-5$ [Hertz]. All the other Solar Planets are similarly located between the point of the maximum Ambient Power and the zero point at the beginning of the accelerated rotation. This unusual Natural Phenomenon is shown in Fig. 3.

18.) Figure 3

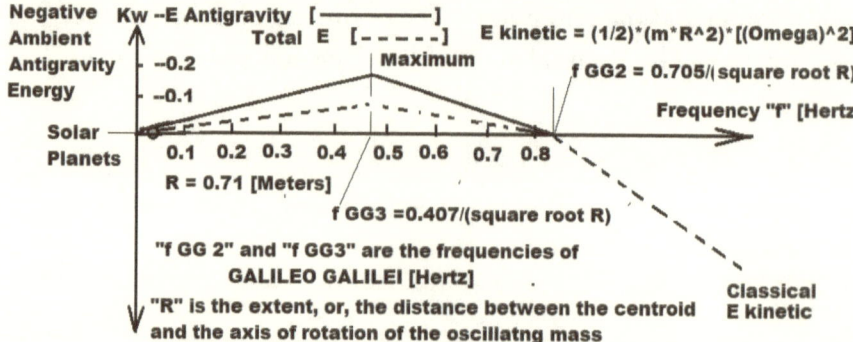

Fig. 3

PREDOMINATING ANTIGRAVITY COMPONENT OF THE AMBIENT ENERGY FOR VARYING AN INCREASING FREQUENCY--WATER PUMP EXAMPLE

"f GG 2" and "f GG3" are the frequencies of GALILEO GALILEI [Hertz]

"R" is the extent, or, the distance between the centroid and the axis of rotation of the osciliatng mass

REFERENCE TO *SPACE ENIGMA*

19.) Antigravity Energy Variance For All The Solar Planets, Exhibiting Three Maximums, In Light Of The Observed Activity At The Sun

The component to be analyzed is the Antigravity Energy, or, the expression 58 (in *Space Enigma, 1.3.3.10*): E Ambient = m*r*g, for rotating mass like spheres throughout the Solar System:

E Antigravity Energy = E Ambient Energy =

= (g^2) *(r^3)/(Gamma) = (Gravitational mass) *(g)*(Planetary Radius "r")

Where:

"r": is the outer radius of the Planetary Shell [Meters]

"+/--g": is the Planetary Surface acceleration [Meters/sec^2]

"Gamma": is the Constant of Universal Gravitation of Sir Isaak Newton.

Analytical results:

Ambient Energy E Ambient = $[(g^2) *(R^3)]/(\text{Gamma})$

The SUN	--3.770*10^41	1.418*10^11	'ln
Planet E A		= Relative to Mercury E A'	
Mercury	--2.658*10^30	1.0	0.0
Venus	--2.874*10^32	108.126	4.68
Earth	--3.713*10^32	144.587	4.97
Mars	--7.587*10^30	2.854	1.05
Jupiter	--3.412*10^36	1,283,671.934	14.06
Saturn	--4.002*10^35	150,564.334	11.92
Uranus	--1.082*10^34	4,070.729	8.31
Neptune	--2.364*10^34	8,893.905	9.09
Pluto	--7.169*10^30	2.697	0.99

"Space Enigma", By Steven L. Basic, Page No. 1.3.3.10

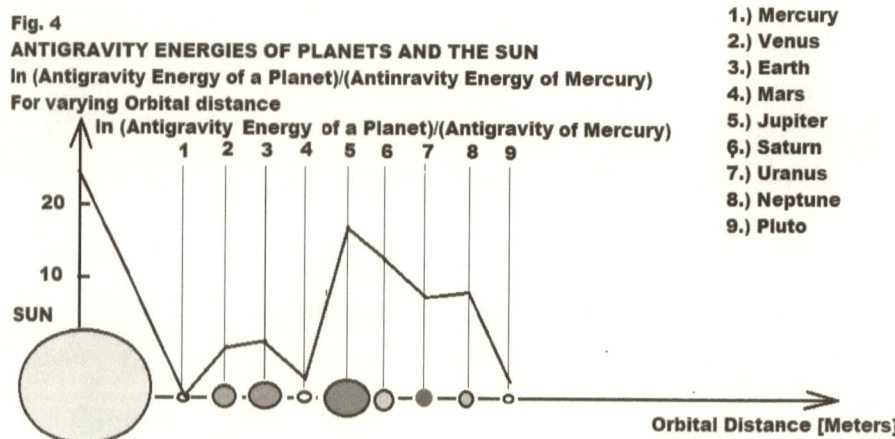

Fig. 4
ANTIGRAVITY ENERGIES OF PLANETS AND THE SUN
In (Antigravity Energy of a Planet)/(Antinravity Energy of Mercury)
For varying Orbital distance

1.) Mercury
2.) Venus
3.) Earth
4.) Mars
5.) Jupiter
6.) Saturn
7.) Uranus
8.) Neptune
9.) Pluto

20.) Tesla Ambient Energy

The relationship between the energy and light velocity squared, the Author has studied to a great extent. In the outer space, the ratio between the Energy and the velocity of light squared holds an immense importance:

$$[\text{Energy}/ C^2]$$

Professor Albert Einstein, and his predecessors (C. Maxwell, H. Lorentz etc.), as supported by the references at the university—level textbooks describing the Mossbauer effect, (the Wikipedia, Encyclopedia Britannica, and more,) all lead to conclusion that this ratio:

Dim $[\text{Energy}/C^2]$ = $[(\text{Kg—of—force}) * \text{SEC}^2/\text{Meter}]$

is very fundamental and important; its dimension appear the same as that of the Mass.

In continuation of this great effort, also at the beginning of the 21st Century, the square of the Mass—like event has been found to be:

$[\text{Energy}/C^2]$ ^2 = ZERO POINT ENERGY = $(h*f)/2$ or, equal to the "ZERO POINT ENERGY".

The foundations for the "ZERO POINT ENERGY" were defined in 1913 by the scientists Einstein—Stern.

THE ENERGY OF A MASS—LIKE--EVENT, APPEARED TO BE PRPORTIONAL TO THE SQUARE OF THE VELOCITY OF LIGHT IN OUTER SPACE, AND THE SQUARE ROOT FROM THE "ZERO POINT ENERGY" OF PROFESSOR EINSTEIN—STERN, AT THE BEGINNING OF THE TWENTY FIRST CENTURY:

$$E\,mass - like - event = (C^2) * \sqrt{\frac{h*f}{2}}$$

BY VIRTUE OF THE SQARE ROOT LOCATION (IN RESPECT TO THE "C^2"), IT HAS BEEN POSSIBLE TO CONCLUDE THAT THE MAIN SIGNIFICANCE OF THE SQUARE ROOT FROM THE "ZERO POINT ENERGY" OF EINSTEIN--STERN, IS IDENTICAL TO A MASS OF THE ENERGY:

$$\left(\frac{E\,nergy}{C^2}\right) = \sqrt{\frac{h*f}{2}}$$

$$= m\,(mass\,of\,the\,Energy\,C^\wedge 2 * m)$$

Hence the Theorem:

$$(m)^2 = \frac{h*f}{2} \qquad \text{And}$$

$$E\,nergy = (C^2) * \sqrt{\frac{h*f}{2}} \qquad \dots\dots\dots\dots\dots\dots\,[1]$$

The expression No. 1, from the previous page No. 46 , can be written:

$$Energy = \sqrt{\frac{C^4 * h}{2}} * \sqrt{f} \quad\text{............................[2]}$$

Where, the parametric variables "C" and "h" in metric [MKS] system are:

1.) The velocity of light in outer space C= 2.9979*10^8 [Meters/second]
2.) Max Planck constant h = 6.4942*10^--33 [Kg*Meter*second]
3.) Dim [f] = 1/sec = Hertz
Substituting "C" and "h" values into the expression No. 2 , an Ambient Energy becomes:

$$Energy = 5.121 * \sqrt{f} \quad\text{............................[3]}$$

The expression No. 3 will be Denoted as the "Ambient energy of Nikola Tesla", that is a wave and, in the same time, an intensity of a vector.

Nikola Tesla has studied the intermediate and high frequencies during very many years of his life. This Energy has been found to be a component of the Scalar Energies.

AMBIENT--IMMATERIAL ENERGIES OF NIKOLA TESLA FOR VARYING MID—RANGE--SPECTRUM FREQUENCIES

Col/Row	1	2	3
	MID—RANGE DESCRIPTION	MID—RANGE FREQUENCY "f"	Nikola Tesla Immaterial-- Ambient Energy= = 5.121* (f ^1/2)
Dim.	"f"	Hertz	Kg*Meter
1	Micro-Pulsation	0.10	2
2	Extremely Low	10	16
3	City Grid	60	39
4	Very—low	10^4	512
5	Television—Radio	360*10^6	97164
6	Microwave	10^9	161940
7	Radar	10^10	512100
8	Experimental	10^11	1619402
9	Infra-Red	10^13	16194023
10	Visible--Light	10^14	51210000
11	Ultra—Violet Light	10^16	512100000
12	"X" Rays	10^18	5121000000
13	Gamma--Rays	10^20	5.121*10^10

21) Addendum

LOW ENERGY NUCLEAR REACTIONS

IN LIGHT OF THE HISTORIC "SCALAR WAVE ENERGY"

EXPERIMENTS OF NIKOLA TESLA

INTRODUCTION

AT THE BEGINNING OF THE 21ST CENTURY, THE EXTENT OF THE THEORETICAL AND EXPERIMENTAL ACTIVITY IN THIS FIELD OF SCIENCE, WITHIN THE UNITED STATES AND WORLD-WIDE, WAS IMMENSE.

THE AUTHOR MADE SEVERAL ATTEMPTS TO ALERT THE SCIENTIFIC CIRCLES CONCERNING THE SUBJECT OF "SCALAR WAVE ENERGY" THEORY FOR OPTIMIZATION, THAT RESULTED IN NO RESPONSE.

IT IS KNOWN THAT NIKOLA TESLA, RESEARCH SCIENTIST AND PHILOSOPHER, IN THE 20th CENTURY, HAS ORIGINATED A NEW "POLYPHASE" ELECTROMOTOR AND AN POLYPHASE ALTERNATOR.

THERE ARE MILLIONS OF THEM BUILT ALL AROUND THE WORLD. HOWEVER, MOST IMPORTANT HIS WORK

HAS BEEN IN THE FIELD OF THE HIGH FRQUENCY ELECTRICITY AND SCALAR ENERGY WAVES.

THE FOLLOWING NUMERCAL EXAMPLE COULD BE FOUND TO BE SUFFICIENTLY ACCURATE FOR THE CIVIL APPLICATIONS, SUCH AS THE AGRICULTURE, TRANSPORTATION, HEATING AND Medical field hospitals.

THE SCALAR WAVE ENERGY INCORPORATES THREE COMPONENTS:

Component 1

"AVOGADRO" NUMBER OF ATOMS, FOR A KNOWN VOLUMETRC INEFFICIENCY OF THE 'LOW ENEGY NUCLEAR RECTOR" CORE, BECAUSE, ONLY TRANSMUTING BOUNDARY LAYER OF THE CORE REMAINS EFFECTIVE, AND BECAUSE THE SCALAR PRODUCT BETWEEN COMPONENTS 2 AND 3 WOULD BE PER ONE SIGLE ATOM.

Component 2

TESLA WAVE AMBIENT ENERGY AT A SPECIFIC FREQUENCY.

Component 3

SYNCHRONISED--MODULATED ON--TO THE TERMAL GENERATOR OF THE ZERO POINT ENERGY—EINSTEIN--STERN— WITH THE TESLA WAVE AMBIENT ENERGY GENERATOR, PRE—HEATED TO THE INFRA RED FREQUENCY.

"AVOGADRO" NUMBER OF ATOMS"

Dia. No. 1

"Low Energy Nuclear Reactor"
   Total Core Volume
Metric [MKS] to Imperial Conversios
1 Micron = $3.937*10^{-5}$ inches; 1 inch = 25400 Microns; 0.050 inches =
= 1270 Microns

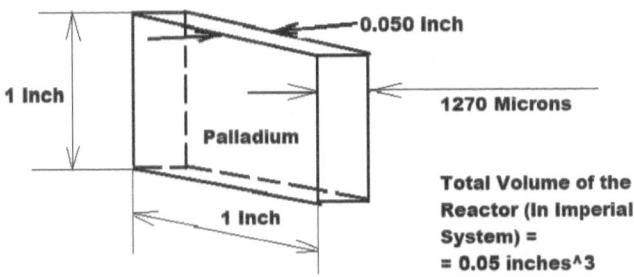

0.050 Inch
1 Inch
1270 Microns
Palladium
1 Inch
Total Volume of the Reactor (In Imperial System) =
= 0.05 inches^3

Dia. No. 2

'Low Energy Nuclear Reactor" Volume of the Transmuting boundary layer
             (20 Microns, both surfaces)
Metric [MKS] to Imperial Conversios
1 Micron = $3.937*10^{-5}$ inches; 1 inch = 25400 Microns;
= (20+20) Microns = $1.5748*10^{-3}$ Inches

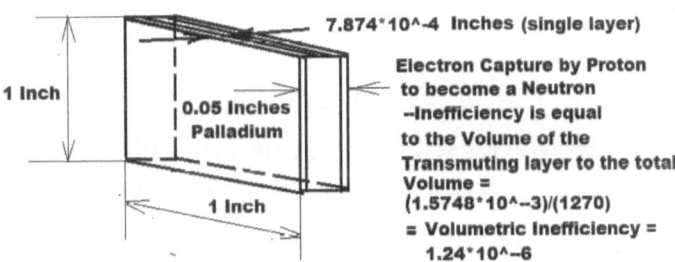

$7.874*10^{-4}$ Inches (single layer)
1 Inch
0.05 Inches Palladium
1 Inch
Electron Capture by Proton to become a Neutron
--Inefficiency is equal to the Volume of the Transmuting layer to the total Volume =
$(1.5748*10^{-3})/(1270)$
= Volumetric Inefficiency =
$1.24*10^{-6}$

"AVOGADRO" NUMBER OF ATOMS, PER ANALYTHICAL CHEMIST SCI CONSULTSNTS OF COLORADO

Avogadro Number of Atoms = (CORE WEIGHT IN GRAMS) *(VOLUMETRIC INEFFICIENCT OF THE CORE

From the page No. 3) *(6.022*10^23)/ (ATOMIC WEIGHT OF THE CORE MATERIAL, FROM THE PERIODIC ELEMENTS TABLE =

Avogadro Number of Atoms =

= (50) *(1.24*10^--6) *(6.022*10^23)/106.4 = 3.509*10^17

TESLA WAVE AMBIENT ENERGY AT THE INFRA-RED FREQUENCY = 1.734 *10^12 [Hertz]

TESLA AMBIENT ENERGY =5.121*(1.734*10^12) ^0.5 =

E 2 = 6743406 [Kg of force*Meter]

SYNCHRONIZED--MODULATED ON--TO THE TERMAL GENERATOR OF THE ZERO POINT ENERGY—EINSTEIN--STERN— WITH THE TESLA WAVE AMBIENT ENERGY GENERATOR PRE— HEATED TO THE INFRA RED FREQUENCY

"ZERO POINT ENERGY" OF EINSTEIN-STERN = (f*h)/2 [Kilogram of Force*Meter]

WHERE "f" IS THE INFRA—RED FREQUENCY =

= 1.734*10^12 Hertz

AND "h" THE MAX PLANCK CONSTANT, IN METRIC [MKS] SYSTEM = $(6.4942*10^{-33})$ [Kilograms of Force*Meter*Second].

HENCE,
E 3 = $(1.734*10^{12}) *(6.4942*10^{-34})/2 = 5.63*10^{-21}$ [Kilogram of force*Meter]

HENCE THE SCALAR ENERGY:
E scalar =
(Avogadro Number of the Core atoms) *(E 2) *(E 3) =
= $(3.509*10^{17}) *(6743406) *(5.63*10^{-21}) = 13322$
[Kilograms of force*Meters]
This Power may be NON—QUANTUM, therefore it will be converted into the KILOWATTS, by dividing E scalar, by "75" and "1.37":

P scalar = (13322) / [(75) *(1.37)] = 129.655 KILOWATTS, per one single square inch, or, two megawatts per square meter.

This power magnitude has been found almost the same as an estimate of the contemporary EXPARMENTAL and THEORETICAL Research: "FOUR MEGAWATTS PER SQUARE METER" (Referred to THE EXCELLENT chronology "Hacking the Atom" by Steven B. Krivit, on amazon.com, THIS MAGNIUDE = FOUR MEGAWATTS PER SQUARE METER will be used to re—calculate the Avogadro number to replace estimated in this section, and the Power per one square meter of the reactor:

Avogadro/Avogadro re—calculated = 3.329

This means that the re--calculated Number of Atoms will be somewhat smaller for this thinner core.

The "Inefficiency" of the Core—signifies a statistical possibility of the deuterium protons, to capture, within the transmuting boundary layer an electron and to became a Neutron.

Hence the Scalar Power becomes:

Power = (Avogadro re—calculated) *(1.662*10^--32) *(f^3/2), or, for Infra-Red frequency f = 1.734*10^12 [Hertz]

P = (1.054*10^17) *(1.662*10^--32) *(2.283*10^18) = 4000. —

Kilowatts = 4.0--Megawatts.

CONCERNING THE NATURE OF SCALAR WAVE AMBIENT ENERGY

"While experimenting with violently abrupt direct current electrical charges, I have found that a new form of energy (scalar) came through..." this Secret Remained eternally with Nikola Tesla.

The Scalar Energy incorporates two waves of identical Frequency from two opposite directions.

A concentration of the (Scalar Wave) Energy Resonates from the Union of two Waves, that is also termed as "Wave Coupling".

Scalar Power (Kilowatts)
$P = $ (Avogadro Number of Atoms) $*(f^{3/2}) *(1.662*10^{-32})$
Where, $(1.662*10^{-32}) = $ Constant and "f" is the Frequency of (Radio) Waves [Hertz]
LOW ENERGY NUCLEAR REACTION, an example:
$P = $ (Avogadro Number of Atoms) $*f^{3/2}) *(1.662*10^{-32})$, or, for Infra-Red frequency $f = 1.734*10^{12}$ [Hertz]

$P = (1.054*10^{17}) *(2.283*10^{18}) *(1.662*10^{-32}) = 4000.$

Kilowatts = 4.0 Megawatts/per square Meter.

22.) References

[1] ACADEMICIAN VELJKO MILKOVIC, 2000, "TWO STAGE OSCILLATORS" MACHINE FOR THE EXTRACTION OF FUEL—LESS AMBIENT POWER FROM THE GRAVITATIONAL STRESS FIELD.

[2] Mr. BRUCE FLETENBERG, OHIO, U.S.A., 18,000 LB WEGHT MACHINE FOR THE EXTRACTION OF FUEL—LESS AMBIENT POWER FROM THE GRAVITATIONAL STRESS FIELD.

[3] Basic, Steven L. "SOME REMARKS CONCERNING THE DISTRIBUTION OF GRAVITATIONAL ACCELERATION AND GRAVITATIONAL MASS OF PLANETS THROUHOUT THE SOLAR SYSTEM". *INDIAN JOURNAL OF THEORETICAL PHYSICS.* VOL. 47, No. 1, 1999

ISSN 0019—5693

Steven L. Basic, Principal Engineer, Boeing Corporation,

P.O. Box—780919, Wichita, KS 67278—0919, U.S.A.

(Received for publication in August 1997)

[4] Basic, Steven L. "CONCERNING THE SIZES OF PLANETS AND THE DISTRIBUTION OF GRAVITATIONAL ACCELERATION AND GRAVITATIONAL MASS THROUGHOUT THE SOLAR SYSTEM. *INDIAN JOUNAL OF THEORETICAL PHYSICS*. MARCH 14, 2001.

[5] The Initial publication: "SOME REMARKS CONCERNING THE GENERAL SOLUTIONS OF THE LEONHARD EULER'S AND JOSEPH LAGRANGE'S DIFFERENTIAL EQUATIONS WITH VARIABLE COEFFICIENTS" By Steven Basic, Senior Aerospace Fatigue Stress Engineer, Supersonic Aircraft Works, British Aircraft Corporation, Cambridge, England.

(Report in Spring of 1968)

CONTRIBUTING REFERENCES:

[6] Albert Einstein and Otto Stern, Analen Der Physik, "On Molecular motion near absolute Zero" (1913)

[7] Basic, Steven L. "Zero Point Energy per Stereo Radian and the Distribution of Gravitational Acceleration of Planets Throughout the Solar System." Library of Congress Control Number: 2013908608. Softcover 978-1-4836-3914-7. E-book 978-1-4836-3916-1. Xlibris.

[8] Dissertation of Madame Marie Curie.

[9] *Mossbauer, R.L. (1958) "Kernresonanz—fluoroscenz von gammastraslung IIV Ir 191, Zeishrift fur Physik* 151(2) 121—143.

[10] Basic, Steven L. Previously published "UNIFIED THEORY" ISBN 978—1—4349—9649—7, Page No. 18, 2009,

[11] Max Planck, Quantum Theory (1901)

[12] Louis De Broglie, Matter—Wave Theory (1926)

[13] Heinrich Hertz, Frequency (1887)

[14] Nikola Tesla (1856—1943)

www.ingramcontent.com/pod-product-compliance
Lightning Source LLC
Chambersburg PA
CBHW031542210526
45464CB00003B/1110